Bedtime Thoughts

Excerpts from
The Power of Control Thought

Roy F. Messier

iUniverse, Inc.
Bloomington

Bedtime Thoughts
Excerpts from The Power of Control Thought

iUniverse books may be ordered through booksellers or by contacting:

iUniverse
1663 Liberty Drive
Bloomington, IN 47403
www.iuniverse.com
1-800-Authors (1-800-288-4677)

ISBN: 978-1-4620-8225-4 (sc)
ISBN: 978-1-4620-8222-3 (hc)
ISBN: 978-1-4620-8224-7 (e)

Printed in the United States of America

iUniverse rev. date: 4/13/2012

About the Author

Roy F. Messier, the tenth of eleven children, was born in Worcester, Massachusetts. After completing school, he served in the Navy for four years and then moved to New York City, where he worked as a hairstylist for twenty years.

Later he studied Culinary Art in California and worked in the restaurant business while studying to become a minister. At the Universal Truth Center of California, he studied medical and metaphysical sciences. After completing Science of Mind courses, he served as a practitioner at the Religious Science Church of the Desert in Palm Desert, California, under Minister Dr. Tom Costa.

Currently residing in California, Mr. Messier paints and writes.

Contents

Introduction

Yes, the Power of Control Thought is within everyone's grasp, although it does take time to master. It's controlling all thought that you have about everything in your world. Change your life from where you are now to where you want to be with the Power of Control Thought. The Power of controlling your thoughts is most important in today's world. Without it, you would live life not fully free from the world around you. The Power of Thought is an energy that, once mastered, can produce a life full of wonders, a life with control – your control, not someone else's thoughts about how you should live.

The Power of Thought is so great that man, even in today's world of great technology, has been able to master only a small percentage of the Power. When you master the Power of Thought, your world will be like a dream of wonderful experiences. With each new day, the unfolding of your life will prove that you are in full control. In any situation you can choose to have thoughts that will govern your environment. When people around you are in disarray, you don't have to be a part of it.

With the Power of Thought, you are in control of what you choose. In the Power of Control Thoughts, you will find that when confronted with anger or unhappiness or any other situations, you, and you alone can be the master of what you do. In this way, only the good comes from your thoughts; you can become involved in the happening or you can walk away from it. You need choose it only if you wish to be part of it. The choice is yours.

It was the study of Religious Science that completely changed my life. It is the foundation for inner peace which is the teaching for everyone's life. One of the most important things that I learned in my studies was the ability to forgive. It is a God-given power that enables me to wipe the slate clean and clear my Mind of debris accumulated in daily living. I have forgiven myself and others. I do not hold any thoughts of unforgiving-ness. Few experiences are as rewarding as those that result from forgiving fully and completely. I know that my state of Mind is vital to my health and wholeness. I now think Control Thoughts are constructive and uplifting. I have learned any change in life comes from negative to positive. Science of Mind is based entirely upon the supposition that we are surrounded by a Universal Mind, into which we think. This Mind, in its original state, fills all space.

ON EVERY PAGE OF THIS BOOK IS A MESSAGE OF HELP.

You can open the book to any page and you'll find the message and help you need at that moment.

Acknowledgements

I would like to express my love, gratitude and thanks to the following people for their valuable assistance in making this book a reality.

My deepest gratitude goes to my special friend André Yan for all his help and support.

To my sister Julie Manzi, my sincere appreciation and loving thanks for all she has done through the years.

Heartfelt thanks also go to dear friends Donald Beck, Geoffrey Webb, Ken and Wendy Anderson, my sisters Emily Prieskorn and Rita Titus and my brother Felix Messier. Additional thanks to my trusted editor, Alaina Bixon.

What is Control Thought?

It is learning to listen to what kind of thoughts we are having and controlling them. Are they positive or negative? When our Mind accepts an idea as being true, it then becomes true for us. Every idea through thought fed into the Mind is bound to produce an effect exactly like its cause. We have the power within to make anything happen in our life.

Control Thought

Can do anything that the mind can receive from the One Mind, that Infinite Intelligence. Thoughts create everything that we experience in life from the very beginning of our life. Our thoughts are the creative avenue of the Mind within us.

With Control Thought, you will see order replace disorder, unity replace separation.

You only have to choose and the Law of Mind honors your selection. Therefore in Control Thought, make wise and loving choices in consciousness, then the unfolding of corresponding conditions will manifest.

One needs to work with Control Thought to create the kind of life one wants to experience.

The thoughts you want to think about are as hard to control as the thoughts you do not want to think about. Our Mind is compelled to create conditions in our life based upon the ideas that are fed into our Mind through thought. Controlling your thoughts can change your life.

We must run our own subjective and emotional life.

If not, the world will run it for us and will impregnate it with all sorts of ideas of limitation. Every person must learn to run his or her own life. For no one can take our place when death comes.

There is one Power.

That which is within. One true Law and that is your own Spirit. It is the only immutable Power we know, our intuition and inspiration, all comes from the direction in which we turn our thoughts.

It is because of man's nature and not his will that his thoughts are creative. Truth is the law of nature at work.

Being clear about this is important. What you think is what you are instructing the Mind to create for you through you—what you can and cannot do. Faith makes what your thought images are, turning them into what you are. Everything in the Universe is in accord with the law, just as faith is a law and acts as such.

know your own thoughts.

Train yourself to think what you wish to think, be what you wish to be, feel what you wish to feel, without limitation. In a similar manner, think right thoughts until your thoughts become perfectly clear.

Everything is from within out.

Even if you have heard this a hundred times, it is good to keep reminding yourself that whenever you move your mind, everything else in your life will start to move in the same direction.

Everything we do is in response to a desire.

Our faith directs us in whatever we do. Faith is a condition of the Mind. With desire and faith, there is a facility called imagination. Imagination is our plan-making department and is under our direction. There is no limit to our ability to use imagination.

Right decision is the action of a Control Thought.

Its being is the consciousness of the One Mind. A person can decide today to reject or accept what he has. One must learn to rule one's life.

The clearer your thoughts are about your real nature, the more you will be able to experience it. We are individuals and must express the individuality to the fullest extent in our personality.

Many people are besieged by unruly emotions.

Feelings may often seem to overwhelm you and tear you apart. "I can't help the way I feel." But is this true? Are you to be lifted to the sky with flights of good feelings only to be plunged low by down-spells? Are you not able to use your intellect to help you recognize your powerful emotions, think them through and change them if necessary?

We are surrounded by a Universal Mind into which we think.

This Mind in its original state fills all space. It fills the space that man uses in the Universe. It is in man as well as outside him. As man thinks, he sets a law in motion which is creative. Every thought in the life of an individual may express itself.

The wonderful thing about the Mind is that we can change our input to create a different result.

Sounds simple, doesn't it? It is.

If your belief systems have been held in place for years, what you need to do is program a whole new input and be faithful about practicing your new patterns of thought.

The change we need to make occurs first within ourselves.

If you are not experiencing the good you desire, you need to change your thinking. For instance, you do not like what you are hearing in your mind, or you are not getting what you are looking for. Then the time is now to change all thoughts that keep you from getting what you want in life.

We all have the ability to choose and decide

All the belief systems that were taught to you in your early days were someone else's thoughts of how they perceived the world. As you learn to grow in the spiritual world, your belief system gives way to a new belief system. This means through your power to choose and decide you can build the personality and character you choose to build. These are all mental facilities and spiritual qualities to be used, expressed and experienced as you grow in understanding yourself.

Cause and Effect

The law of cause and effect is very simple and very sure in its working. Cause and effect are inseparable. You should think of them as always being together.

Thought is the cause. Action is the effect.

We can change any effect by changing the cause.

Think what you want to happen. Expect your thinking to come to pass. Learn to be specific. This gives your creative power direction. Cause and effect are inseparable and you should think of them as always being together.

Always remember that all you can ever do is support or attempt to support each other. You can only engage in unity or division.

The law of cause and effect operates on our beliefs as we actually believe them to be.

Each of you can take charge of your own life in a way that will lead you to experience a deep sense of joy, satisfaction and fulfillment. It all starts with the awareness that the Mind can and does control all that you are from the very beginning.

It is a universal law: if we reverse the action of a cause, we can at the same time reverse the effect.

Negative thoughts produce negative results. Positive thoughts produce positive results.

We have the power within to make anything happen in our life.

You have the same power to change any situation you created out of the thoughts you fed into the mind.

What are your basic attitudes toward life?

This is an important question for you to answer since your attitudes affect your life so profoundly. For instance, happy people tend to have a healthy attitude of self acceptance, feeling that they deserve to be peaceful and happy; while those who are not happy tend to display the attitude that they are not deserving of life's good.

With the strong belief we
hold about ourselves, we
cannot fail to assume shape
and form in the outer world.

Redirect your thinking until you gain new and more positive beliefs about who you really are. Let go of all of those negative thoughts you have of yourself and bring into your thoughts only the positive ones.

There is a difference between faith and belief.

One can have faith with little belief. But a belief in faith reaches a point that you no longer have to work at it. It is always there in your thoughts. Like thoughts of "I can" instead of thoughts of "I cannot."

Every time we think, we are
thinking into a receptive
plastic substance, which
receives the impression
of our thoughts.

When you stop to realize how subtle thoughts are, how
unconsciously you think something negative, how easy
it is to get down and out, mentally, you see that it is
perpetuating its own condition.

Why is this happening to me?

Life is your attempt to integrate what you see happening with what you think should be happening. You are spending time asking God to explain to you the insanity of the world you yourself created, through your attitudes and beliefs. The real question is show us the blessings or the lessons you would have us learn. God's voice is everywhere. Accept the answers. Be patient.

Control Thought establishes us to greater faith and good.

Right thoughts govern your affairs, for you have complete dominion over your thoughts, in charge of whatever goes on in your life. What do you think you can or cannot do? Thinking clearly about this is important. What you think is what you are instructing the Mind to create for you.

Listen for reassurance

The right answers are never forced upon you. They require your willingness to want them. Then it is given to you or in reality you become aware of them. After you have taken them in, the ones that will work start to grow and become apparent.

Our inner guide will gladly affirm that our thoughts are true.

The reason is simple. They are in concert with our highest mind. Our highest self or mind is always in concert with God, the source of all knowledge.

Man is a spirit while God is The Spirit.

Within the infinite Mind, each individual exists not separated but as a separate entity. You are not separated from life but are a separate entity in it. You are individualized in your own world.

The secret of getting in shape mentally is to clear our negative beliefs and limiting ideas.

When you do so, you reshape your destiny. Let's stretch your mental muscle and reach into a living-ness you have only imagined up to now. Be filled with enthusiasm and zest for mentally shaping up. Release all negative ideas from life now and replace them with new vitalizing, energizing thoughts of love, life and power.

There is One Infinite Mind.

This is from which all things come. This Mind is through, in and around man. It is the only Mind there is and every time man thinks, he uses it. There is one limitless life which returns to the thinker exactly what he thinks into it.

Move from the impossible to the I am Possible.

Become enthusiastic over new creative ideas and continue to learn and grow. In this manner, you will remain young in heart and your body will reflect your thinking at all times.

Man is an individualized center of Divine Thought.

Through him, the Original Thinker is finding a fresh starting point for its creative power. Creative Minds are one with the Universal Intelligence.

All habits are objectified thoughts.

You are not a victim of the thoughts, opinions, theories or beliefs of others; nor your own personal fears. Our own subconscious is amenable to change. That is its nature.

Take stock of the thoughts you are entertaining in your mind.

If there are any that do not promote success, let go of them and replace them with Control Thoughts that produce wholeness and happiness. This is a good day to start weeding out unwanted thoughts and planting Control Thoughts that you wish to see flourish.

We are the conductor of our orchestra of life.

Do you ever notice others playing the wrong notes of negative ideas, such as:

I don't feel well;
Things are a mess; or, I can't stand my neighbors!!
Is it any wonder that life seems to be less than harmonious?

Think about it, maestro.

I will be master of my own actions.

I will never permit myself to become confused or perplexed. A powerful Will gives a person the ability to emit powerful energy into a given thought and keep it there until the goal is attained.

We are all receptive to the larger Mind, of which we are all a part.

Its power is your power. Its wisdom is your wisdom. Its nature is your nature. It has the availability and the willingness to be the creative impulse in your thoughts. Control Thought has the ability to cause wonderful and rewarding experiences in your life.

The Mind is the creative principle of life.

Each person has access to it and each person has an instrument which can be used to bring into life whatever it wants. When you know how to control your thoughts, then each one of you can map his or her own destiny without limitations. This instrument is the Mind.

The energy of mind, like other natural energies, already exists.

You should carefully consider whether you are willing to experience the results of your thoughts. If you keep your thoughts fixed upon the idea that this energy, which is also Intelligence, is now taking form of some desire in your life, then it will begin to take that form. The thoughts that you concentrate on become your attention.

We are dealing with a Universal Principle through our thought power.

Why should you set any limit to its Power? Universal Principle is the law of the Universe. The law knows nothing about limitations.

Our reason for being is Universal Life, which is our origin.

Truth is the door to freedom and knowing that you are perfect just as you are. Live in peace and know that harmony is your every experience in life. Things are guaranteed to work for you if you keep your attitude correct.

Our experiences are the direct result of our own state of mind and nothing else.

You are the final arbiter of your own fate. When our mind accepts an idea as being true, it then becomes true for us. Every idea through thought fed into the mind is bound to produce an effect exactly like its cause. We have the power within to make anything happen in our life. Also, we have the same power to change any situation we create out of the thoughts we feed into the Mind.

*Guilt and fear cause
some ideas and beliefs
to be forced below the
conscious level, where they
are hidden away from us.*

When you are willing to seek out these separated thoughts
and have them exposed, they will be brought to truth
where they can be seen as illusions. They will return to
the nothingness which they really are. In reality, you are
always guided and cared for while remaining in constant
communication with your source of knowledge, power
and harmony.

All that we become and do and hear in the physical life is prepared behind the veil within us.

Therefore, it is of immense importance in life to grow conscious of what goes on within these domains. You must be the master. You must be able to feel, know and deal with the secret forces that determine your destiny and your internal and external growth or decline.

Acknowledge the One Spirit, the one mind of the Universe

In everything you do, there is a logical sequence of plan, proceed and prosper. Thoughts need to be followed by action. You prosper as a result. You have the capacity to follow through on anything you want to create, including a new way of living. Life is always ready, willing and able to work through you.

Have the acknowledgement and faith in knowing there is a power within

There is overwhelming evidence that this power does exist. The basic teaching of religion is that there is an infinite power in the universe. This power is accessible to all of us if we believe though faith. Since the power in the universe is everywhere, this presence is in and through us. This presence manifests itself in and through all forms, all people and all conditions.

"Be not conformed to this world, but be transformed by the renewing of your mind."

Devote each day to get yourself into a regular routine of thinking Control Thoughts. You have to practice an attitude of patience. With patience, all Control Thoughts will be mastered. Faith and belief make you a believer.

The Universal Law of Mind translates our desires into reality.

This law is the reflection of your own thoughts. Awareness of this possibility provides you with a powerful tool when the human condition yearns to reach a high level, throwing off the pain and loneliness of separation and experiencing unity with others. In meditation, you connect with the One Power and in doing so, you are all mentally connected.

Everything in the Universe emanates a Vibration.

Thought atmosphere is what I call thought vibrations. As a rule, people make use of and think thoughts received from the thought waves of others.

Mental atmosphere influences the power of attraction.

It is almost entirely subjective: you meet someone, only to turn away, while we are drawn to others without any apparent reason. This is the result of mental atmosphere or thought vibration. Each person has a mental atmosphere which is the result of all that one has thought, said and done.

LIFE as we know it is a form of energy and has to be directed by man's intelligence.

Our body is filled with super atomic power. Our human body is composed of trillions of atoms and all have a center of intelligence. There is a new belief that there are brain cells within each cell throughout our physical being, which means that there is intelligence throughout our body.

In the universe of our mind, we are at the very center of the Universe

Know that this divine presence and power is always with you. Acknowledge the presence that is the life within you at this very moment. Give expression to this power within.

What the mind can conceive and believe, the mind can achieve.

Visualize your intended destination. Your subconscious Mind is affected by this self-suggestion. You must remember whenever your subconscious Mind accepts an idea, it immediately begins to execute it. This law is true for good or bad ideas.

*To win one has to begin
to set into action that
which needs to change*

Expanding Spiritual Awareness

Conscious Mind—our five senses.

Subjective Mind stores our memory

Subconscious Mind—The Mind of God.

As you grow in spiritual awareness, you grow in the understanding of God within. Nothing is impossible. Only in our state of mind can we imagine it as being impossible. The result from your acts may vary, but it is possible to visualize something that is possible.

Life is a mirror and will reflect back to the thinker what he thinks into it.

You should never allow yourself to think of or talk about limitations or poverty. To view limitation is to impress it upon the Mind. Limitations and poverty are not things but are the result of restricted ways of thinking.

Believe in yourself. Believe in your ability to give yourself successfully to whatever is important to you, and know that this is true.

To aid in bringing this about, you must fire the imagination with an idea and believe totally in your ability to achieve it. Turn on the power of enthusiasm in regard to whatever your desires may be. If you are not experiencing the good you desire, you need to change your thinking.

The realization of your perfection is the most inspired goal you could ever have.

You already know how creative you are. Your thoughts are always at work fashioning the events that shape your days. Through our negative thoughts, we sometimes create failure, sickness and low self-esteem. You must create an image of wholeness and perfection and not spend your precious time immersed in thoughts of limitations and frustrations.

Yes, life is a NOW EXPERIENCE. There is a law that responds to whatever we are thinking.

The more spiritual the thought, the higher its manifestation. Spiritual thought means an absolute belief in and reliance upon truth. There is no other. There is no opposite and there is no truth whatsoever to lesser ideas. There is One Infinite Mind from which all things come.

Universal Truth is a principle of law.

The truth is that without love, law is incomplete. The two are one. Everything works in the Universe by the Law and love being all, the two are one. In using the law of your being with this comprehension of self, physical expression is available, active and abundantly endowed in wholeness.

Realize that it is a law of reflection.

Everyone has a personality and can use it to the fullest if one is with the Infinite Mind—one with the personality of God.

If you want a friend—be a friend. The power of attraction emanates from within. Whatever you reflect into Mind tends to take form.

All our thoughts are to bring our consciousness to spiritual awareness

Pure spirit is at the center of every organ, action and function of man's being. You are using a principle which automatically reacts to you by corresponding with your mental attitudes. Your spiritual awareness is the secret place of the most high within you. Everything is created out of yourself from all of your thoughts.

We are not separated from life, but are a separate entity in it.

If the ideas of lack are your dominant mold, Mind has no alternative but to produce lack. Conversely if the ideas of abundance are dominant, the Mind will act on this idea and produce those things that represent abundance to you.

Each one of us is life personalized.

You are living life as a person. Therefore, each of you contains within all the intelligence and power you need. Each person has the ability and the power at their command to express life in peace, happiness, abundance and satisfaction.

An attitude that is destructive to our happiness is the belief that we do not compare favorably with others.

Such a feeling of inadequacy deprives you of the freedom to move joyously forward in life. However, there is a wonderful way to offset this attitude—by radiating love and goodwill to everyone around you. Your attitude toward life and other people will be transformed into a lighter dimension of thinking and being.

To be self-conscious is to be a spiritual entity.

Spirit is first cause—the conscious mind and the power which knows itself as a conscious being. You live and move in your own world controlled by the thoughts you put into your mind.

Our consciousness is established in right ideas and entirely supportive emotionally, which operates as the law of our life.

Here and now recognizes that everything about you that really matters is indestructible and inseparable from you. It is your capacity to understand that you are fully alive and fully aware that what you really are is Mind thinking clearly.

We have a vast subconscious power at our disposal.

The thoughts you want to think about are as hard to control as the thoughts you do not want to think about. For example, you can think yourself into being happy or unhappy. Your mind is compelled to create conditions in your life based upon the ideas that are fed into your mind through thoughts. In controlling your thoughts, your life will change.

Logic and reason in the subconscious mind mean nothing.

Logic can be the greatest deterrent to a successful life. Using logic, you may ask yourself, "Am I worthy of these goals or am I really entitled to them?" Whatever your goals are, you can have them with or without logic.

The creative intelligence of our subconscious mind knows what is best for us.

Its tendency is always lifeward and it reveals to you the right decision which blesses you and all concerned. You will recognize the answer when it comes; it has all to do with your consciousness.

Our body and all its organs were created by the Infinite Intelligence in our subconscious mind.

It knows how to heal you. Its wisdom fashioned all your organs, tissues, muscles and bones. This infinite healing process within you is now transforming every atom of your being, making you whole and perfect now.

We must stay inside the
house, so to speak, and
allow one wiser than we
are to guard the door.

Then all ideas that should come in, do so. Those that do
not belong go away or simply exit through the back door.
You must, however, stand guard over your own thoughts
and encourage your thoughts to be loving, open, trusting
and honest.

The energy of any belief we
have compels us to behave
in a manner that is in
harmony with our beliefs.

In the most remarkable manner, you are responsible for
creating the way you feel. You are responsible for your health
and your well-being. You are responsible for the money you
have and successes you enjoy. You are responsible for the
nature and quality of all your relationships with others.
You are responsible for everything with no exceptions.

We sense that behind the word which we speak is the power of the Universe surging to express itself.

It is good to become an observer of your thoughts, to see which thoughts are based on pre-judgment, fear or lack of trust, and which thoughts are trusting, open, loving and peaceful. You must define your thoughts as to what words you will speak. Labels cause pain. To judge another is really hiding your self. Words you use, in a very real sense, determine your belief system.

We must take action
immediately. Stand up
and speak the word of
peace to the storms
of human thought.

Always remember that all you can ever do is to support
or attempt to suppress each other. You can only engage in
unity or division. Inevitably, whatever you promote, you
reproduce.

Before Self is the doorway for change

So whatever it is you wish to experience in your life, hold the thought. Have the experience fill your heart and mind with thoughts of these experiences, and that is exactly what you will create. Be here Now, for it is there you will have a new beginning, a new life. Make the commitment to yourself to make every moment count. For the Now is all we have.

Spiritual mind practice is an uncovering of the Divine nature in us.

What the Divine has implanted cannot be uprooted. It can only appear to be covered up. Always, on the other side of confusion, there is peace. Always on the other side of discord is poise. The Universe can only give you what you are aware of. So the spiritual mind is the creative factor. Everything is created out of yourself from all of your thoughts.

All worthwhile messages are of the Spirit; their form is unimportant.

Don't worry about the form nor about the person who delivers it. The messenger can come from a Willie Nelson song such as "Live One Day at A Time," or from a five year old child, a religious leader, a spiritual writing, a politician, the "Daily News," a billboard or our worst enemy or a dream.

All things are echoes of God's voice; Be open to all sources.

The truth will stay and the rest will pass by. You need to let everything come in and to have your inner guide do the judging. The result is echoes of God's voice. Be open to all sources.

It is good to remember that inner peace begins with a thought.

Be open to good, for you cannot harbor negative thoughts and expect good to flow from you. You are part of the One Mind, the Intelligence of the Universe, but at the same time, you think as an individual.

The person who lives spiritually
does not live much differently
from others, except he
has fewer problems.

There is a way to begin now to become more aware of the spirit in yourself and in others. Your positive thoughts and ideas direct you into positive action and deeds.

know that the principle of Faith is always in action through the law of mind

The absolute truth is God and nothing else. The absolute is a way of living, thinking and believing. Know that this principle of Faith is always in action through the law of Mind at the level of God.

Faith is a position of confidences, assurances and rest in the mind which comes from listening to the true voice of God within.

The higher the sense of truth, the greater will be the realization. The highest is the power of the thought with the unity of the spirit. The action of thought with spirit produces creation.

You are a part of Infinite Life, which knows no end.

You are a child of eternity. You are wonderful. You are greater than any opinion, conscious or subconscious that you may hold about yourself. Recognize that opinions, at best, are preliminary knowledge leading to truth. Intelligence information is a tool which Mind within you uses, pre-existing within you daily to improve your thoughts.

The source of true happiness and joy is always within you.

Know that your happiness does not depend on outer conditions or persons. See all your experiences in life as moments in the evolving individual experience of the whole self.

All our peace and contentment comes from within.

Everything is accomplished through life and life is of the Mind and in the Mind. Life is forever present in its fullness. Right here and now is the only place you can find peace and you only have to go within.

Retirement is a new venture

It is a new challenge, a new path and the beginning of the fulfillment of a long dream. Look at it this way—retirement is a promotion from kindergarten to the first grade. The ladder of life is education and the understanding of life in general. Retirement is still another step forward on the ladder of life and that step is the gaining of wisdom. Retirement is not the end but the beginning to live life freely, mentally and physically.

Thought by Thought

What does it mean? It means you can stop playing at life and start living it. Everything in our life is totally controlled by our thoughts. Control Thought allows you to change your thoughts that will produce the life you want.

I choose to think positive thoughts about myself and my environment.

What you experience today is the direct result of what you have chosen, either consciously or subconsciously in the past. You can re-choose, therefore, and create what you want to experience now.

You created yourself

You stand at the doorway of your mind and you alone can select the thoughts that make you what you are. Give thanks for your creativity.

Life is a continuous process of growth.

You move from classroom to classroom. You do not stay the same as you were ten years ago or even yesterday. Changes happen to you every day and you grow from these changes every day.

Daily devotion; all efforts made in this regard will be rewarded.

In reality, there is only one guide, one source, and it can be experienced in many forms. It is helpful to set aside certain times during each day devoted just to the purpose of getting in touch with your inner guide. The messages from your inner guide are always of love and support.

The creative power is the thinker behind our thoughts.

You need to have the willingness to allow change to take place. Clearing the path for newness in your thoughts, you have to be aware that the first thing to do is clarify your thoughts. Changing them into a wonderful demonstration of your own elevated thoughts, which reveals through your awareness who and what you really are.

All is love. Love points the way and the law makes the way possible.

When you come to the understanding that all is love and yet all is law, the conscious use of understanding as it operates from the Mind affects the entire body. The starting point is in the realization of understanding that the One designated as the Almighty is a presence dwelling in you; a force surrounding you and a principle by which you live.

Love breaks down the
iron bars of thought
and shatters the walls
of false belief

You must remember above all that the soul of life is love, and that love shows itself by service. Service proceeds from sympathy, which is the capacity for seeing things from the point of view of those whom you would help, while at the same time seeing them in their true relation.

Hate, transformed to love, sees everyone through the eyes of love.

Starting today, see everyone through the eyes of love, allowing this powerful force to guide your thoughts, guard your words and direct your activities. When negative and contrary thoughts surface, and they occasionally do, inwardly say no to those thoughts. Tell them to go away, that they are not true and not welcome.

Love is a divine energy that begins in God and has no end.

Love is not an emotion that begins in you and ends in the positive response of another. I believe that when you have love in your heart for all, then you are motivated by love from within and then empowered by it. Approach others from the most loving position each day, always desiring good communication as you relate.

Love is the center of man's being.

The Law of Mind acts upon the ideas you carry in your heart, in the deepest recesses of yourself. Those ideas which you accept and believe are what the laws act upon, causing them to be realized in your experiences.

Love is a mental as well as a physical state.

You know that to the degree that you perfect your thoughts, then the perfection of all men will appear to you. For unconditioned love lights the pathway of life. One of the most important things for you to remember is that you are always causing something to be created for you. For the energy of thought through the mind is always producing the universal law. The law allows the thought, the thought follows desire.

Everyone deserves to have love.

Regardless of where you may be on your pathway of learning, release into the world of experience the expanded sense of life and love that has been absorbed into your consciousness.

Love and appreciate yourself, that uniqueness that you are.

Choose to experience your limitless potential. Choose to make the most of yourself and to live your life fully and richly. Go about the business of doing something about the changes that are needed for growing and enriching your life.

The source of true happiness and joy is always within you.

Know that happiness is not the result of acquiring any particular thing such as wealth or status; it is the result of recognizing your true nature. That which is within. Know that your happiness does not depend on outer conditions or persons. Start now to recognize and accept the truth for yourself.

Our aim in life is to give wonderful service and all those whom we contact are blessed by what we have to offer.

Creative wisdom works through you bringing all your plans and purposes to completion. Whatever you start, you bring to a successful conclusion. All your work comes to full fruition in divine order.

The language of tone is the language of the spheres.

It is the language of the universal world. A mental message can be sent to reach a certain person or it can be sent so that it reaches a great number of people. This tone may, however, be raised or lowered by choice, which is why the mental organization of different people usually has different tones, and yet there are many people in the world with exactly the same tone, and they all vibrate in unison.

It is a laboratory of individual unfolding.

Marriage can only succeed when both parties see something of the divine in one another. True love is spiritual perception. Undoubtedly, there will be challenges or tests for you in many areas of life.

A peaceful state of mind is a healing state of mind.

You must decree that you are under the influence of Divine understanding and Divine Intelligence. There is no condition that cannot be overcome. Hold to the conviction. Believe it, affirm it. Know that the more we hold to the spiritual idea of health, the stronger we become in faith. The more we affirm the idea of health, the more gradually we can make a channel for the healing power.

The body is healed as the inner mind is transformed.

When old and false images of negative thoughts are renewed by positive thoughts, the Mind is transformed, and healing can take place. You must forgive and release all old beliefs. You cannot be healed of any problem you continually look at and constantly describe. The perception of wholeness is the consciousness of healing.

We are tuned into the healing vibrations that our mind is tuned into.

If you continue to think there is no cure for your illness, then there is no cure because you are tuned in on a negative channel. The healing channel cannot get through if your thoughts are in disharmony. You cannot receive anything from vibrations until you are tuned into yourself.

Through Control Thought, you can heal yourself of any belief which gives power opposed to good.

The thoughts you want to think about are as hard to control as the thoughts you do not want to think about. Your mind is compelled to create conditions in your life based upon the ideas that are fed into your mind through your thoughts. Controlling your thoughts can be changed any time in your thinking. Things may happen around you and to you but the only things that really count are the things that happen in you. You cannot always control what happens to you but you can control what your thinking is.

You can be healed, no matter who you are or what you have been or done, if you make contact with that inner life and faith.

When you are conscious of perfect life, the body is whole. You must become unconscious of the imperfect and conscious of the perfect alone. So if you desire to be healthier, then start doing your part to consistently and immediately reverse all negative thoughts of sickness and disease.

Mental healing is the result of clear thinking and logical reasoning which presents itself to consciousness and is acted upon by the mind.

It is a systematic process of reasoning. Thinking sets causation in motion. Right thoughts constantly poured into consciousness will eventually purify it.

*You have within you
the greatest physician
in the world – the
power of the mind.*

The Power of the Mind is what causes your heart to beat, your muscles to strengthen and your body to heal. Praise your body, your mind and your emotions. Feel good about yourself. It is because of man's nature and not his will that his thoughts are creative.

Ponder the miracle of your body.

The miracles, to name a few, performed daily and without complaint, are controlling your heartbeat and respiration, digesting your food and compound chemicals to make you strong, renewing cells and combating disease. These miracles heal wounds and maintain your equilibrium among all the body's other enormous tasks to keep you operating at peak efficiency. Ponder these miracles and count your blessings.

We are a perfect entity, living in a perfect universe.

You must know that the all-powerful Spirit is ever available to the healing of any discordant condition of the body, mind or affairs. To find this Spirit, look within, for it is dwelling in you right now and has always been there from the very beginning.

The more we affirm the idea of health, the more we make a channel for the gradual healing power.

There is no condition that cannot be overcome. Hold to this conviction. Believe it. Affirm it. Know that the more you hold to the spiritual idea of health, the stronger you become in faith. The more we affirm the idea of health, the more we make a channel for the healing power.

The outer rim of reality is exactly at the center of itself.

When your thoughts know freedom, the law will free the body and outer life will express health, happiness and success. It can be found in no other place.

Every day I am becoming more and more lovable and understanding.

This affirmation establishes a major premise in your thinking, that the infinite intelligence of your subconscious mind is guiding, directing and prospering you spiritually, mentally and materially. Another affirmation that will help is the following: I am now becoming the center of my world.

Mental attitude is a thought away.

If you have thoughts of love and you draw upon these thoughts of love, then you receive more love and the more loving you are to yourself and others. If you have thoughts of intelligence and you draw upon these thoughts of intelligence, the more intelligent you become and act.

Everything that lives, lives in light; everything that is in existence, radiates light.

By intuition, you are brought into contact with the inner light. Then the truth is known to you. You come to the understanding that all is love and yet all is Law.

Through the action of law, love and friendship are attracted automatically to us.

People are lonesome because they have a sense of separation from people. The thing to do is not to try to unify with people, but with the principle of life behind all people and things. To do this, you must go within to that center of your life and get in touch with the idea of oneness, beginning to feel at one with all of life as it expresses through all people everywhere. Unify with the idea of friendship to all.

Think of the whole world as your friend; Be a friend of the whole world.

Love is the grandest healing and drawing power on earth. See only good in people. Refuse to let yourself misunderstand or be misunderstood. Know that everyone wants you to have the very best in life.

I am the thinker in my world.

You have to believe that the movement of thought is the movement of Mind. Your part is thought, the other action. You must look upon the process, not as the thing itself, but as a way of arriving. Having control of your thoughts and projecting them into specific conditions will bring the reality of life.

I am that I am, the master of my thoughts in my world.

This is a dynamic action. It is an action of consciousness unfolding. It is this energy that moves through heightened emotion to a conscious growing awareness of truth. You have to believe that the movement of thought is the movement of Mind.

I will be master of my own actions.

I will never permit myself to become confused or excited. That thought may be for some personal action, or for the action of another person, or even for a body of people.

To have peace, one has to make the decision to have Control Thought of peace at all times.

You have heard so often, "I am not at peace." To hear this, then the person has decided wrongly in their thoughts about peace. Decisions cannot be difficult. This is obvious if you realize that you must already have decided not to be wholly joyous if that is how you feel. Therefore, the first step in the undoing is to recognize that you yourself actively decide.

Proceed as if at peace.

You do not have to ask your guide for permission to do everything. This speaks more of fear than of trust. If your peace becomes disturbed, stop and ask for guidance and never force things. You should join your will with God's will. This is letting go.

The starting point to all achievement is definitiveness of purpose.

Set goals that you can expect results from. Goal setting is one of the most exciting and rewarding habits you can become involved in. It is the process which allows you systematically to get exactly what you want.

What is security? It is faith and belief.

Whatever you think you can do or dream you can do, then do it. There is a power within that reaches the goal. You build a mental equivalent for what you want by getting interested in it. This is the way you create feeling. Action leads to success.

Dream your dreams and
then translate them
into organized thoughts.
Action leads to success.

A burning desire to be and to do is the starting point
from which dreams take off. Belief in the power of desire,
backed by faith, will bring success through the powerful
principle of mental Control Thought.

Maintaining a state of mind, known as a burning desire to win, is essential to success.

Wishing will not bring riches, but desiring riches with a state of mind that becomes an obsession and planning a definite way and means to acquire riches is action. The world is full with an abundance of opportunity. Remember, no one is ever defeated until defeat has been accepted as a reality.

Blessed are the men who have found their work.

The spirit of man is the pure spirit of God within. You are atoms of the Universe living in human matter. One of the secrets of a forceful mind is knowing that everything is in order and is being taken care of by the Higher Power.

Vision: see your idea as creative action causing things to happen.

See your source of being. Vision is an important factor in your progress. It is a creative action within the great law of the Universe. In other words, stop dreaming of what you would like to be or do. See your vision as already existing. Only that which you see is included in your experiences. See the perfect you. See the self in your action. What you see in Mind, you see as your objective success and you as a success.

Creative Visualization

The process of creative visualization is to imagine as clearly and realistically as possible what you want to happen, as if it will happen, or is happening. In creative visualization, you need to focus on specified areas of your life, one at a time, such as relationships, financial matters, health and family, to name a few. All these categories may not fit into what you need to do, so create your own.

Creative visualization is a good tool to use for goals.

Picture in your mind a goal that you would like to have. Utilizing the visualization technique is by far the most reliable tool to see a new and more prosperous future for yourself.

Meditation is the breakthrough in consciousness.

It is the awareness of a higher power that can and will guide you through your life. Meditation actually is using life's law and love, this being the power of the spirit within. At which time you will have that inner experience of peace.

A meditation program has one critical element, and that is being realistic in how much time you will spend on meditation.

One of the most important aspects of meditation is the follow through. A good meditation is straightforward for a particular problem. Meditation then is when the mind and spirit become one and in doing so, become the power to change the changeable to the reality of your acceptance.

Meditation is quite similar to a good physical program.

Both require hard work and, like exercise, meditation has no age limits. Training the Mind, the way athletes train their bodies, is one of the primary aims of meditation.

Improper breathing deprives your brain of sufficient oxygen.

This can be caused by preoccupation and stress. Most of us in our busy lives breathe short breaths, from our chest alone, instead of deeply from the abdomen.

Auto suggestion is a principle that works.

Auto suggestion (instructions to the subconscious mind) is a principle that works. Meditation and the use of auto suggestion, through which any person may use to rise to altitudes of achievement, stagger the imagination. Auto suggestion can change what the subconscious has already stored from negative to positive.

Affirmation is the act of affirming a positive assertion.

The best way to do this is by writing out your affirmations and reading them daily. You cannot transform outer matter without transforming inner matter, for its origin is always the same thing.

The one wholly true thought one can hold about the past is that it is not here.

To think about it at all is therefore to think about illusions. The mind is actually blank when it does this because it is not really thinking about anything. You see only your thoughts projected outward. The mind's preoccupation with the past is the cause of misconception about time. Your mind can only grasp the present, which is the only time there is.

Re-program your beliefs.

Become that person you have always wanted to be. You have to remember that your programming determines your beliefs and your beliefs determine what you do and have. Re-programming is basically new thoughts of how you want your life to be: as you change the inner attitude of your mind, your subconscious mind responds.

The value of solitude cannot be overestimated.

All great deeds are born in solitude, and all great characters are formed there. All good impulses are stimulated by judicious solitude and concentration of thoughts.

Today is the first day of your life, so today, recognize and practice the ideas of patience in every area of your life.

Let go of striving to make things happen. Let go and let the power of Control Thought reveal through you the desire of the heart. Accept new ideas with an open mind. Don't worry about what people might say regarding your endeavors. Concern yourself only about the constructive side of life. Patience, indeed, does have its reward in the ultimate realization of success.

Action in Life

Honesty, fairness, caring, respect, loyalty and tolerance.

Action cannot get you anywhere unless you start. Don't forget that people will judge you by your actions, not your intentions.

Begin where you are, but do not stay where you are. The best time to do something worthwhile is between yesterday and tomorrow. What you say and what you do advertises what you are. Failure always overtakes those who have the power to do without the will to act. The only thing you have to fear is not doing something about the fear you have.

Attitude

Attitude is a mental position with regard to a fact or state of Mind. An attitude you have about everything can make your life full of wonderful experiences. The attitude about work, family. self and the people in your world, all of your attitudes are greatly dependent on your thoughts at all times.

Behavior

Some people continue to change jobs, mates, and friends but never think of changing themselves. Behavior is what a man does, what he thinks, feels or believes all through his actions in everyday life.

Education

The mind is like the stomach. It is not how much you put into it that counts, but how much is digested. A hundred mistakes are a liberal education, if you learn something from each one.

Courage

More twins are being born these days than ever before. Maybe kids lack the courage to come into the world alone. Courage is being the only one who knows you are afraid.

Doubt

To become a living embodiment of success, your belief in yourself needs at all times to be so firm that it is not short-circuited by doubt, fear or self-distrust. You have to let go of thoughts of doubt, distrust, worry, condemnation, and fear, replacing them with positive thoughts. When you say, "I will try to do that," by saying, "I'll try," you have already accepted defeat. What you are really saying is either, I don't know if I can or I don't know if I want to. Both of these statements denote doubt, and doubt is a giant obstacle to the manifestation of good.

Ego

Your own state of mind is a good example of how the ego was made. Belief is an ego's function. As long as you are open to the belief, your ego arose from separation and the continued belief in separation. You, who believe in ego, cannot believe in God, for ego is self-serving. Your own power along with your Control Thoughts can guide you to the real relationship that exists between God and his creations.

Enthusiasm

A wise man once said that enthusiasm is nothing but faith with a tin can tied to its tail. Enthusiasm is the propelling force necessary for climbing the ladder of success.

The three F's in Life

Faith, for it is faith that forms your outlook on life. Family, for it is the family that you were born into that you will have while you are here. Friends, for it is the friends along the journey that provide you with the understanding of other humans. All three F's in your life are there to fulfill your life.

Fear

Every attitude contains in it the seed of a corresponding thought or experience. You have many fears in life. To overcome fear wherever it may be, say to yourself slowly, quietly and positively, I am mastering this fear. I am overcoming it now. I am relaxed and at ease and I know that it is so. As these positive seeds of thought sink into the subconscious, they grow after their own kind and you become poised, serene and calm.

Forgiveness

Why is it difficult sometimes to forgive others? Is it because what they did was so unforgivable or because you have become attached to negative memories, through continually thinking about them? Releasing the past is the first step toward complete forgiveness. You can let go when you realize that the words and actions of others are responses from their own beliefs and their responsibilities, not yours. You are growing and unfolding at your own rate, and no two people think or feel in exactly the same way. Knowing this, you release the need to have everything happen the way you think it should.

Guidance

You may ask yourself, am I experiencing the kind of problems that do not seem to have a solution, or do I feel defeated even before I begin searching for an answer because the situation calls for more understanding that I think I have? The answer to these questions and others comes in turning within to that power, Control Thought, aligned with Divine Intelligence and Divine Understanding, which is available to you whenever you need it.

Happiness

You know intuitively that lasting happiness is not dependent on outer sources or conditions. It is only through the recognition, development, and the use of our inherent qualities that you can find true happiness. Only when you develop positive thoughts about yourself, the world around you, and the people in your life. This is the beginning of happiness. Happiness is found within. All the thoughts that you have on being happy will produce the happiness in your outer life.

Harmony

The thoughts of a loving relationship. It is like a blueprint of a home or a recipe for a favorite dish. You know the right materials or ingredients are essential for a successful outcome. In order to live, work or be in harmony with others, you need to include all the right ingredients in your relationships with them. You start with a heaping measure of love. You include a generous portion of faith. You add a pinch of patience to allow the very best results to come about. .

Health

Health is the condition of being sound in Mind. It is freedom from physical disease or pain. Just what is health? Health is not just of the Body but also of the Mind and Spirit. Some of your health problems arise from your thoughts, some from what you eat and drink. Then, there are the health problems from accidents, such as taking too much of the wrong drugs, transportation, and home accidents. All can be worked out in time. Your thoughts about any or all of these problems can be of great help in your healing. Of course, you need to see a doctor, but working with your thoughts can and will bring you back to health.

Idea

An idea is more than information, it is information with legs and it is headed somewhere. Words should be used as tools for communication and not as a substitute for action. You make your future by the best use of the present. Many of us have a tendency to put down our ideas – we tell ourselves this is not a good idea. But a lot of the time, it might be a good idea and you need to recognize it.

Inner Peace

- Tendency to think and act spontaneously rather than from fear based on past experiences.
- An unmistakable ability to enjoy each moment.
- Loss of interest in judging yourself.
- Loss of interest in judging other people.
- Loss of interest in interpreting the action of others.
- Loss of interest in conflicts.
- Loss of ability to worry and to feel guilty.
- An overwhelming appreciation.
- Contented feeling of connectedness with others.
- Frequent attempts to smile through the eyes from the heart.

Leadership

Leadership can be taught. The qualities for leadership can be developed, made available for you through training in self-expression and self-control. An effective leader is a Communicator. The leader's task is to communicate the vision, the mission and the goals to the members. Becoming a more effective communicator will allow you to sell ideas to employers and employees, improve family relations and win new friends. Communication creates meaning for people. It is the only way any group can get behind the goals of an organization. You must impose upon yourself the discipline of seeing things from the viewpoint of others.

Love

Stop searching for love outside yourself. Love as you would be loved. Become aware of those negative thoughts that are not in tune with the Infinite beauty of nature. Everything works in the Universe by law, and love being all, the two are one. You are only at the beginning of an eternal journey, and all along the way you should respond to the highest healing power of all – love. Believe that when you have love in your heart for all, then you are motivated by love. Love keeps open the channels of communication in all relationships.

Revenge

Revenge is a response to a feeling of hurt. It is a very primitive response that allows people to feel they have enacted justice for whatever reason. In a sense, it is not being able to cope with the hurt. On the other side of the pain, there is relief and great peace of mind when you forgive everyone who has caused you pain. In this way, the thoughts you have while thinking of revenge are all negative thoughts. When you can change them to forgiving positive thoughts, you will find you're free of the hurt.

Riches

Start now to think you can and believe you can. Move from the impossible to the "I am" possible. Yes, life is a now experience. There is a law that responds to whatever you are thinking. Maintaining a state of mind known as a burning desire transforms thought into concrete action.

Wishing will not bring riches, but desiring riches with an obsession, and planning definite ways and means to acquire them, will. The world is filled with an abundance of opportunity. Limitation and poverty are not things, but are the results of restricted ways of thinking.

Vision

When you look at a great masterpiece of art, you know that the artist must have held an inner vision of beauty and form while completing it. From the first chip off the marble or the first stroke of the brush on a canvas, the artist was picturing his or her own inner vision. You, too, can be an artist who creates masterpieces in your life. Your inner vision of health can affect your body in positive ways. Your inner vision of peace shows forth as harmony in your relationships. By holding to your inner vision, you are allowing the peace of within to come forward, showing all who you are.

Books of Interest

The Power of the Subconscious Mind by Dr. Joseph Murphy

Wisdom of the Ages and *The Power of Intention* by Dr. Wayne W. Dyer

Living in The Light and *Creative Visualization* by Shakti Gawain

The Seven Spiritual Laws of Success by Deepak Chopra